GALACTIC SOCIETIES

GALACTIC SOCIETIES

CARL NYELL

Contents

1

Introduction to Galactic Societies

The Search for Extraterrestrial Life

The search for extraterrestrial life has captivated humanity for centuries, prompting profound questions about our place in the cosmos. This inquiry encompasses not only the scientific exploration of distant planets and moons but also the cultural implications of discovering intelligent beings beyond Earth. Astrobiologists rigorously study extreme environments on our planet, seeking analogs for life in the harsh conditions of other celestial bodies. From the icy crust of Europa to the sulfuric clouds of Venus, each discovery enhances our understanding of the potential for life elsewhere and fuels the imagination of UFO believers and conspiracy theorists alike.

The scientific community has made significant strides in recent decades, primarily through initiatives like the Kepler Space Telescope and the ongoing Mars missions. These endeavors aim to identify exoplanets within the habitable zone of their stars, where conditions might support life. The discovery of biosignatures—chemical indicators of life—in the atmospheres of these distant planets has become a focal point for astrobiological research. The tantalizing possibility that we may one day intercept signals or discover artifacts from extraterrestrial civilizations has led to a surge in public interest and speculation, creating a fertile ground for conspiracy theories around government secrecy and potential cover-ups.

Extraterrestrial cultures, should they exist, would likely possess unique social structures, languages, and belief systems that could challenge our understanding of existence and consciousness. The diversity of life on Earth suggests that alien civilizations could exhibit a wide range of evolutionary paths, each shaped by their environment and history. This cultural diversity could inspire profound philosophical and ethical discussions about how humanity interacts with other intelligent beings. The implications of such interactions raise questions about cooperation, conflict, and the sharing of knowledge, which are often central themes in conspiracy theories regarding potential government interactions with extraterrestrial entities.

The notion of extraterrestrial ecosystems further complicates our understanding of life beyond Earth. Each ecosystem, shaped by its planet's geology and climate, would dictate the forms of life it could support. The study of extremophiles on Earth—organisms thriving in the most inhospitable conditions—provides insights into the adaptability of life and the potential for diverse biological forms to arise elsewhere. For UFO believers, the existence of such ecosystems supports the argument that intelligent life could be visiting Earth, exploring our planet as part of their own ecological studies or seeking to establish contact.

As the search for extraterrestrial life continues, the intersection of scientific inquiry and public fascination with UFO phenomena highlights a broader cultural narrative. Government disclosure initiatives, aimed at revealing information about unidentified aerial phenomena, have further fueled speculation and conspiracy theories. The blending of rigorous scientific pursuit with the allure of the unknown creates a dynamic discourse about our universe. As researchers and enthusiasts alike grapple with the implications of potential discoveries, the ongoing search for extraterrestrial life remains not only a scientific endeavor but also a profound exploration of what it means to be part of a vast, interconnected cosmos.

Overview of Alien Civilizations

The exploration of alien civilizations encompasses a vast array of cultures, technologies, and ecosystems, each reflecting unique evolutionary paths and sociopolitical structures. These civilizations, shaped by their home planets' environmental conditions, exhibit diverse forms of life and intelligence that challenge our understanding of biology and society. Astrobiologists have long speculated about the potential for life beyond Earth, and the study of these hypothetical civilizations provides a framework for comprehending the myriad ways sentient beings might develop. By examining the potential characteristics of extraterrestrial societies, we gain insights into their possible social norms, governance, and interaction with other species.

In the realm of extraterrestrial species, the variations are as limitless as the cosmos itself. From insectoid beings thriving in high-gravity environments to aquatic civilizations existing beneath the waves of oceanic planets, each species is adapted to its surroundings in ways that science is just beginning to grasp. The physiological and cognitive traits that emerge from these adaptations can lead to vastly different cultural evolutions. Understanding these differences is crucial for UFO believers and conspiracy theorists who seek to understand the implications of contact with alien life. The ability to categorize and theorize about these species assists in forming plausible narratives regarding their potential intentions and capabilities.

Extraterrestrial civilizations may also differ markedly in their technological advancements and cultural practices. Some may have achieved a level of sophistication that allows them to traverse interstellar distances, while others may remain entrenched in their own planetary systems, developing rich cultures without the influence of external forces. The concept of a galactic society hinges on the interaction and exchange between these diverse civilizations. This interplay can lead to the rise of alliances or conflicts, shaped by differing ideologies, resources, and technological capabilities. Such dynamics could mirror historical human interactions, providing a fascinating lens through which to view the potential for cooperation or conflict among alien societies.

The ecosystems of alien worlds further complicate the picture of extraterrestrial civilizations. The unique climatic and geological features of different planets can foster distinct forms of life and societal organization. For instance, civilizations on a planet with extreme weather patterns may develop resilience and adaptability, resulting in a culture that values resourcefulness and innovation. Understanding these ecosystems is essential for astrobiologists who aim to identify habitable worlds and predict the types of life forms that may evolve there. This knowledge also serves as a foundation for conspiracy theorists who speculate about the ecological implications of alien visitation and the potential for contagion or ecological disruption.

Finally, the discourse surrounding government disclosure and conspiracy theories adds another layer to the study of alien civilizations. The secrecy surrounding alleged extraterrestrial encounters fuels speculation about what knowledge has been withheld from the public. This environment creates an atmosphere ripe for the development of theories about the nature of alien civilizations, their intentions, and the broader implications of their existence. The intersection of scientific inquiry and popular belief in extraterrestrial life reflects humanity's deep-seated curiosity and fear regarding the unknown. As our understanding of the universe expands, so too does our capacity to engage with these extraordinary possibilities, calling for a nuanced exploration of the potential for contact with alien civilizations and the cultural ramifications it would entail.

Importance of Understanding Extraterrestrial Cultures

Understanding extraterrestrial cultures is essential for a comprehensive exploration of the cosmos and the myriad forms of life that may inhabit it. As we delve into the possibilities of intelligent life beyond Earth, the significance of comprehending their cultures becomes paramount. This understanding not only enriches our perspective on the universe but also enhances our ability to engage with these potential civilizations in a meaningful way. By recognizing the diversity and complexity of extraterrestrial cultures, we can better prepare ourselves for future encounters and interactions.

The study of extraterrestrial cultures provides insights into the fundamental aspects of existence that may be shared across the cosmos. Just as human cultures are shaped by their environments, histories, and social structures, so too might alien civilizations reflect their unique circumstances. By examining the potential cultural frameworks of extraterrestrial species, we can draw parallels and contrasts that illuminate our own societal norms and values. This comparative analysis fosters a deeper appreciation for the diversity of life in the universe and encourages a more empathetic approach toward beings that may be vastly different from us.

Moreover, understanding extraterrestrial cultures allows us to speculate on their technological developments and social organization. Knowledge of how other civilizations might govern themselves, communicate, and innovate can provide valuable lessons for humanity. For instance, insights into sustainable practices adopted by advanced alien societies could inform our own efforts in addressing environmental challenges. By learning from the successes and failures of extraterrestrial cultures, we can potentially avert similar pitfalls in our development as a species.

The implications of extraterrestrial cultural understanding extend to the realm of governmental policies and public discourse. As discussions surrounding government disclosure of UFO encounters and alien life continue to gain traction, a well-informed populace is crucial. UFO believers and conspiracy theorists often emphasize the need for transparency regarding extraterrestrial phenomena. By fostering a nuanced understanding of alien cultures, we can contribute to more constructive dialogues about the implications of such disclosures, promoting a rational and informed approach rather than one driven solely by speculation or fear.

Finally, the pursuit of knowledge regarding extraterrestrial cultures can play a significant role in shaping our collective identity as inhabitants of the cosmos. As we explore the potential for life beyond Earth, we are confronted with profound questions about our place in the universe and the meaning of existence itself. By striving to understand the cultures of alien civilizations, we embark on a journey that not only expands our horizons but also invites us to reflect on what it means to be human in an ever-expanding universe. This quest for understanding is not just an academic exercise; it is a vital step toward a future where humanity can coexist with other intelligent beings, fostering mutual respect and collaboration across the stars.

2

Extraterrestrial Species

Classification of Alien Species

The classification of alien species is an essential aspect of understanding the broader context of extraterrestrial civilizations and their potential interactions with humanity. In the quest to identify and categorize these species, researchers have proposed various systems based on biological, technological, and sociocultural criteria.

These classifications often reflect the diverse forms that life can take in the universe, encompassing everything from microbial entities to highly advanced civilizations capable of interstellar travel. By analyzing these classifications, UFO believers, conspiracy theorists, and astrobiologists can gain a more nuanced perspective on the potential for life beyond Earth.

One of the most widely discussed frameworks for classifying alien species is the Kardashev Scale, which categorizes civilizations based on their energy consumption. Type I civilizations harness energy at the planetary level, Type II at the stellar level, and Type III at the galactic level. This classification provides insight into the technological capabilities and societal structures of alien species, suggesting that more advanced civilizations may possess knowledge and technology far beyond our current understanding. Such classifications not only highlight the vast potential of intelligent life forms but also raise questions about their motivations, governance, and interactions with less advanced societies.

Another approach to classification focuses on biological characteristics, drawing parallels with terrestrial life forms. This system includes categories such as carbon-based life, silicon-based life, and even hypothetical forms of life that utilize alternative biochemistries. The diversity of potential biological forms challenges our conventional understanding of what constitutes life and opens the door to a wide range of extraterrestrial ecosystems. By studying these classifications, astrobiologists can better predict the types of environments that may support alien life and the characteristics of species that might evolve within those ecosystems.

In addition to biological and technological classifications, the sociocultural dimensions of alien species are crucial for understanding their interactions with one another and with humanity. Different civilizations may have distinct social structures, belief systems, and cultural practices, all of which influence their approach to diplomacy, conflict, and cooperation. This sociocultural classification can help UFO believers and conspiracy theorists frame their understanding of reported encounters and sightings, as it emphasizes the complexity and diversity of potential extraterrestrial societies.

Finally, the implications of alien species classification extend into the realm of government disclosure and conspiracy theories. The potential existence of advanced civilizations raises questions about the extent of knowledge held by governments regarding extraterrestrial encounters. The secrecy surrounding UFO sightings and alleged contact with alien species fuels speculation about the motivations for withholding information from the public. Understanding the classification of alien species can provide context for these theories, suggesting that the nature of these civilizations may influence governmental responses and the broader societal impact of disclosure efforts.

Notable Species in UFO Encounters

Notable species reported in UFO encounters often serve as focal points for both intrigue and speculation within the ufology community. These species, frequently described by witnesses, present a variety of physical attributes, behaviors, and purported technologies that fuel discussions about extraterrestrial civilizations. Accounts range from humanoid forms to more exotic life-forms, each contributing to a broader understanding of what intelligent life might exist beyond Earth. Among these, the Greys, Reptilians, and Nordics stand out, often mentioned in connection with abductions, governmental cover-ups, and the search for truth surrounding extraterrestrial interactions.

The Greys, characterized by their small stature, large heads, and minimal facial features, dominate many UFO encounter narratives. Witnesses frequently describe them as having a clinical demeanor, engaging in procedures that suggest advanced biological knowledge. These beings are often associated with abduction scenarios, where they reportedly conduct experiments on human subjects. The prevalence of Greys in popular culture and conspiracy theories has led to a widespread belief that they are part of a larger extraterrestrial ecosystem, possibly serving as agents of a more complex interstellar society.

In contrast, the Reptilians present an entirely different profile, often depicted as tall, scaly beings resembling reptiles. Their appearances are frequently linked to theories about ancient civilizations and the interconnectedness of human history with extraterrestrial influences. Proponents of these theories suggest that Reptilians may hold positions of power on Earth, manipulating global events from behind the scenes. This has led to a rich narrative surrounding their culture, purported hierarchies, and the implications of their presence in human affairs, all of which provoke fervent debate among conspiracy theorists.

The Nordics, often described as tall, blonde, and blue-eyed, present an image that is strikingly similar to typical Western ideals of human beauty. Reports suggest that they exhibit a more benevolent approach compared to their counterparts, often described as advocates for peace and environmental preservation. This characterization has led to theories about their motivations, suggesting that they may be visiting Earth to guide humanity toward a more sustainable future.

Such narratives align with the aspirations of astrobiologists who ponder the implications of extraterrestrial contact on human development and ecological stewardship.

The exploration of these notable species in UFO encounters not only enriches the tapestry of extraterrestrial lore but also raises critical questions regarding the nature of interspecies interactions. As believers, theorists, and scientists analyze these accounts, they must consider the intersection of cultural beliefs, psychological phenomena, and potential realities of life beyond Earth. Understanding the narratives surrounding these species could offer insights into how humanity perceives its place in the universe and the potential for future contact with alien civilizations. As the discourse around government disclosure and the quest for truth continues to evolve, the examination of these notable species remains a vital component in unraveling the complexities of extraterrestrial cultures.

Biological Diversity Across the Galaxy

Biological diversity across the galaxy presents a complex tapestry of life forms and ecosystems that challenge our understanding of biology and evolution. As astrobiologists continue to explore the possibility of extraterrestrial life, the variations in environmental conditions, planetary compositions, and evolutionary pathways suggest that life may manifest in forms vastly different from those on Earth. The immense variety of potential habitats, from the frigid surfaces of icy moons to the scorching atmospheres of gas giants, implies that unique biological adaptations are not only probable but expected. These diverse life forms may range from microscopic organisms to complex, intelligent civilizations, each adapted to thrive in their respective environments.

Extraterrestrial ecosystems could exhibit intricate interdependencies akin to Earth's own biospheres, though the fundamental principles of life might diverge significantly. For instance, some theorists speculate that life on planets with high sulfur content could utilize biochemistry based on sulfur rather than carbon. Such organisms might possess entirely different metabolic pathways, enabling them to harness energy from their surroundings in ways that are foreign to terrestrial life. The study of extremophiles on Earth, which survive in extreme conditions, provides a crucial foundation for understanding how life could potentially adapt to the harsh environments of other celestial bodies.

The possibility of intelligent extraterrestrial civilizations raises profound questions about cultural diversity across the galaxy. If life exists elsewhere, it likely encompasses a broad spectrum of societal structures, belief systems, and technological advancements. The interactions between these civilizations could mirror the complex dynamics of human societies on Earth, with varying degrees of cooperation, competition, and conflict. The exploration of these potential cultural differences invites speculation about communication methods, artistic expressions, and social hierarchies that might emerge in alien societies, shaping their unique identities in the cosmic landscape.

Government disclosure and conspiracy theories surrounding extraterrestrial life often intertwine with the scientific community's findings, fueling public interest and skepticism alike. As reports of unidentified aerial phenomena (UAPs) gain traction, the call for transparency from governments worldwide has intensified. Many UFO believers and conspiracy theorists argue that there is a wealth of undisclosed information regarding extraterrestrial encounters and technologies. These narratives often serve as a backdrop for discussions about the implications of contact with alien civilizations, including the ethical considerations of how humanity might engage with intelligent life forms.

In conclusion, the exploration of biological diversity across the galaxy is not merely an academic pursuit but a reflection of humanity's quest for knowledge and understanding of our place in the universe. As we gather more evidence through scientific inquiry and technological advancements, the potential discovery of extraterrestrial life will undoubtedly reshape our perspectives on biology, culture, and existence itself. The synthesis of scientific exploration, cultural curiosity, and the intrigue surrounding government secrecy will continue to fuel the dialogue between astrobiologists, UFO enthusiasts, and conspiracy theorists, driving us closer to unraveling the mysteries of life beyond Earth.

3

Extraterrestrial Civilizations and Cultures

Social Structures of Alien Societies

The social structures of alien societies present an intricate web of interactions that can vary dramatically from our own. In examining these structures, it becomes evident that they may be influenced by a range of factors including environmental conditions, biological imperatives, and the technological advancements of the species in question. These influences often lead to the formation of unique hierarchies, governance systems, and social norms that reflect the values and needs of each civilization. For UFO believers and conspiracy theorists, understanding these variables can shed light on the potential interactions and relationships between humans and extraterrestrial beings.

One significant aspect of alien social structures is the diversity in governance models. Some species may adopt a communal approach, where decisions are made collectively, reflecting a deeply ingrained cultural emphasis on cooperation and shared responsibility. Others might have more autocratic systems, perhaps driven by a need for efficiency or survival in harsh environments. Astrobiologists studying the evolutionary paths of these species might discover that their governance styles are not merely social constructs but are instead rooted in their biological makeup. The implications of these governance structures are profound, particularly in how they might influence the behavior of these civilizations in contact with humanity.

Additionally, the role of technology in shaping social structures cannot be overstated. Advanced civilizations may leverage technology to enhance their social organization, enabling more effective communication, resource distribution, and conflict resolution. For instance, a species that has developed sophisticated AI could operate under a technocratic system where decisions are based on data-driven insights rather than human emotions. Such technological integration poses intriguing questions for conspiracy theorists who speculate about the potential implications of such systems if they were to interact with human society.

Cultural practices within alien societies also reflect their social structures. Rituals, traditions, and even art can serve as social cohesion mechanisms, reinforcing group identity and shared values. These cultural elements may serve as a means of communication, both within the species and in potential interactions with humans. Understanding these cultural dimensions is essential for those interested in the broader implications of extraterrestrial contact, as they provide insight into how these civilizations may perceive and engage with humanity.

Finally, the study of social structures in alien societies raises important questions about the nature of cooperation and conflict across the galaxy. As different civilizations encounter one another, their social frameworks will inevitably influence their ability to collaborate or engage in conflict. This dynamic is particularly relevant for UFO believers and conspiracy theorists, who often speculate on the motivations behind reported encounters and the potential for both peaceful exchanges and hostile confrontations. By analyzing the social structures of these alien civilizations, we can gain a deeper understanding of the complexities involved in interstellar relations and the potential outcomes of humanity's quest for contact with the cosmos.

Language and Communication Among Species

Language and communication are fundamental aspects of any intelligent society, shaping the way individuals interact within their species and with others. In the context of extraterrestrial civilizations, the complexity of communication systems can be as diverse as the species themselves. Understanding the nuances of alien languages involves not only deciphering their verbal and non-verbal cues but also appreciating the cultural and contextual factors that influence their communication styles. This complexity raises intriguing questions about the potential for interspecies communication, particularly between humanity and extraterrestrial beings.

Various forms of communication can be observed in the animal kingdom on Earth, ranging from vocalizations to body language and chemical signaling. These methods suggest that advanced civilizations might employ similarly sophisticated systems. For instance, certain species utilize intricate sounds to convey information about their environment, social structure, and emotional states. By extrapolating these observations to potential extraterrestrial species, we can theorize that alien civilizations may utilize a combination of auditory, visual, and possibly even telepathic signals to interact among themselves and with other species.

Moreover, the concept of language transcends mere words and symbols. It encapsulates the shared understandings and cultural values of a species. For extraterrestrial civilizations, language might incorporate elements such as music, light patterns, or other sensory stimuli that reflect their unique experiences and environments. This diversity emphasizes the need for astrobiologists and linguists to adopt interdisciplinary approaches when attempting to decode alien communication, as traditional linguistic frameworks may prove inadequate in understanding the full spectrum of extraterrestrial interactions.

In the realm of governmental disclosure and conspiracy theories, the topic of alien communication has sparked numerous debates and speculations. Many UFO believers assert that there have been encounters with extraterrestrial beings who possess advanced communication abilities, suggesting that these civilizations may already be attempting to make contact with humanity. The implications of such communication are profound, raising ethical questions about how we would respond to messages from other intelligent life forms and what responsibilities we hold in fostering understanding across species.

Ultimately, exploring the intricacies of language and communication among species not only enriches our understanding of potential extraterrestrial cultures but also challenges our perceptions of intelligence and interaction. As we continue to investigate the cosmos in search of life beyond our planet, the principles of communication will undoubtedly play a crucial role in shaping how we engage with and comprehend these alien civilizations. The quest for understanding language among species is not just an academic pursuit; it is a journey that may redefine our place in the universe and our relationship with other intelligent beings.

Beliefs and Religions of Extraterrestrial Cultures

The exploration of beliefs and religions among extraterrestrial cultures presents a fascinating dimension in understanding the broader implications of intelligent life beyond Earth. As humanity delves into the cosmos, the potential for encountering alien civilizations raises profound questions about their worldviews, values, and spiritual practices. Speculative frameworks suggest that just as human societies have developed diverse belief systems shaped by their environments and histories, so too might extraterrestrial societies possess unique philosophies that influence their social structures and interactions with the universe.

In examining the potential of alien religious systems, one must consider the environmental contexts that could shape an extraterrestrial civilization's beliefs. For instance, a species originating from a planet with extreme conditions may develop a reverence for resilience and adaptability, possibly manifesting in deities or mythologies centered around survival and transformation.

Similarly, societies thriving in resource-rich environments might cultivate a belief system centered on abundance and stewardship of their ecosystems. Such variations underline the significance of ecological factors in shaping the spiritual narratives of alien cultures.

Furthermore, the methods of communication and cognition among extraterrestrial species could fundamentally alter their religious expressions. For instance, species with a collective consciousness may experience spirituality not as individual beliefs but as a shared awareness of existence, leading to a communal approach to worship and morality. This could contrast sharply with the individualistic spiritual practices observed in many human cultures. Understanding these variations in consciousness and communication is essential for astrobiologists and theorists as they attempt to hypothesize about the intricacies of alien belief systems.

Additionally, the intersection of technology and spirituality in extraterrestrial cultures warrants attention. Advanced civilizations may integrate their technological prowess into their religious practices, potentially viewing technology as a conduit to the divine. This fusion could manifest in rituals that involve the manipulation of energy or matter, blurring the lines between the sacred and the scientific. Such perspectives challenge the traditional human dichotomy of faith and reason, prompting UFO believers and conspiracy theorists to rethink the implications of technology in the context of extraterrestrial life.

Finally, the implications of governmental disclosure regarding extraterrestrial encounters may lead to a reevaluation of human beliefs and religions. As evidence of alien civilizations potentially surfaces, it could catalyze a paradigm shift in how humanity perceives its place in the universe. This could prompt an unprecedented blending of spiritual beliefs, fostering an era of interstellar dialogue that honors the diversity of thought across civilizations. Understanding the beliefs and religions of extraterrestrial cultures not only enriches our grasp of the cosmos but also encourages a deeper contemplation of our own humanity in the face of the unknown.

Art and Expression in Alien Civilizations

Art and expression in alien civilizations serve as critical windows into the psyches and values of non-human societies. Just as human art reflects cultural narratives and societal norms, the artistic endeavors of extraterrestrial species can reveal their unique perspectives on existence, emotion, and the cosmos. These forms of expression may manifest through various mediums, including visual art, music, performance, and even digital creations, each offering insights into the complexities of their cultures. As we explore the art of alien civilizations, we must consider how their environments, biological makeup, and social structures influence their creative expressions.

The environmental context plays a significant role in shaping the art forms of extraterrestrial species. For instance, a civilization thriving in a gas giant may develop art that utilizes floating materials or luminescent gases, creating a dynamic interplay of light and color in their surroundings. In contrast, a species inhabiting a rocky, mineral-rich planet might focus on sculptural forms using the abundant resources available to them. This adaptation to their environment not only affects their artistic choices but also their worldview, highlighting the interconnectedness of culture and ecology in the universe.

Furthermore, the biological characteristics of alien species can lead to entirely novel forms of expression. For example, a species with advanced bio-luminescence might communicate emotions and stories through intricate patterns of light emitted from their bodies, creating a living art form that changes with mood and context. Similarly, a civilization with heightened auditory capabilities may develop complex musical traditions that resonate with the very fabric of their world. These unique artistic expressions underscore the diversity of life and creativity across the galaxy, challenging our preconceived notions of what art can be.

Just as human art often serves political and social purposes, alien civilizations may utilize their artistic expressions to convey messages about governance, identity, and community. Art may function as a tool for social cohesion, a means of protest, or a method of preserving history and tradition. Ritualistic performances could serve to strengthen communal bonds, while visual story-telling may articulate the struggles and triumphs of a society. By examining these artistic expressions, we can gain a deeper understanding of the social dynamics and values that underlie extraterrestrial cultures.

In the context of government disclosure and conspiracy theories, the study of alien art and expression can also provide fertile ground for speculation. Believers in extraterrestrial life often point to unexplained phenomena and artifacts as evidence of contact with advanced civilizations. The analysis of purported alien art could bolster claims of encounters with otherworldly beings, suggesting that these expressions might hold keys to understanding their intentions and messages. As we delve into the artistic realms of alien societies, we open the door to not only appreciating their creativity but also unraveling the mysteries of their existence and our potential connections to them.

4

Extraterrestrial Ecosystems

Planetary Environments and Their Inhabitants

The diversity of planetary environments throughout the galaxy plays a crucial role in shaping the civilizations and cultures of their inhabitants. Each planet presents a unique combination of atmospheric conditions, geological features, and climate variations that directly influence the development of life. From the harsh, arid deserts of a tidally locked world to the lush, bioluminescent forests of a water-rich moon, the ecological niches available to extraterrestrial species are as varied as the environment itself. These ecosystems not only dictate the biological forms that evolve but also the social structures, technologies, and belief systems of the civilizations that arise.

Consider the potential for life on gas giants, where floating organisms might thrive in the thick atmosphere, utilizing buoyancy to access energy from the planet's storms. Such beings could possess advanced forms of communication, relying on atmospheric vibrations rather than sound. Their societies may be organized around the complex weather patterns, with cultural practices tied to the rhythms of their gaseous world. This notion challenges our terrestrial understanding of civilization, suggesting that intelligence could emerge in forms vastly different from those found on Earth.

In stark contrast, rocky planets with extreme temperatures might support life in subterranean habitats, where geothermal activity provides warmth and energy. These underground civilizations could develop intricate networks of tunnels and chambers, leading to a unique cultural identity centered on resilience and adaptation. Their technologies might revolve around the manipulation of minerals and heat, resulting in a society that values craftsmanship and resourcefulness. The study of such environments expands our understanding of the potential for life and the myriad ways in which it can manifest.

The implications of these diverse planetary environments extend into the realm of government disclosure and conspiracy theories. As astrobiologists and enthusiasts alike ponder the existence of extraterrestrial civilizations, the question arises: how much does our own government know about potential contact with these beings? The variability of life across the universe raises the stakes for transparency. If advanced civilizations exist, understanding their environmental contexts could be pivotal in making contact or even establishing diplomatic relations.

Ultimately, the exploration of planetary environments and their inhabitants not only fuels the curiosity of UFO believers and conspiracy theorists but also provides a framework for understanding the broader implications of life beyond Earth. By examining the diverse ecosystems and the civilizations they foster, we can begin to appreciate the complexity of the universe and the potential for interstellar connections. As we push the boundaries of our knowledge, the quest for understanding these alien cultures will continue to inspire both scientific inquiry and imaginative speculation.

Interactions Between Species and Their Ecosystems

Interactions between species and their ecosystems are crucial for understanding the dynamics of life beyond Earth. In examining extraterrestrial species and their civilizations, it is essential to consider how these organisms adapt to their unique environments and the relationships they forge with other life forms. The study of these interactions not only broadens our perspective on life in the universe but also raises profound questions about the nature of intelligence, cooperation, and competition among diverse species.

Extraterrestrial ecosystems may vary dramatically from those on Earth, influenced by factors such as planetary composition, atmospheric conditions, and available resources. For instance, a planet with a dense atmosphere may support life forms that rely on gaseous exchanges and exhibit adaptations to utilize atmospheric compounds for energy. These unique environmental factors dictate the types of organisms that can thrive and shape the intricate web of interactions within the ecosystem. Understanding these relationships can provide insights into the potential for societal development among extraterrestrial civilizations, revealing how ecological pressures can influence cultural evolution.

The interactions between species can be categorized into various forms, including symbiosis, predation, and competition. Symbiotic relationships, where two or more species benefit from their association, may be particularly prevalent in extraterrestrial ecosystems, fostering a cooperative spirit among intelligent life forms. Conversely, predatory dynamics may lead to the development of complex behaviors and strategies for survival, pushing species to evolve in response to one another. The balance of these interactions ultimately shapes the cultural narratives of extraterrestrial societies, influencing their technological advancements and social structures.

Moreover, the potential for interspecies communication and cooperation raises important questions about the nature of intelligence itself. If extraterrestrial civilizations are capable of forming alliances and sharing knowledge across species lines, it could lead to a more interconnected galactic society. This prospect invites speculation about the role of government entities, such as those implicated in conspiracy theories regarding UFOs and extraterrestrial contact. The interactions between various civilizations may not only impact their ecosystems but also challenge the established narratives of human understanding regarding life beyond Earth.

In conclusion, the study of interactions between species and their ecosystems offers valuable insights into the complexities of extraterrestrial life. These relationships are fundamental to the development of civilizations and cultures, influencing their technological and social evolution. As we continue to explore the possibilities of life in the cosmos, it is essential to consider how these interactions may illuminate the broader tapestry of existence and our place within it. Understanding these dynamics could ultimately reshape our perceptions of intelligence, cooperation, and the interconnectedness of all life forms across the universe.

Adaptations to Unique Extraterrestrial Conditions

The exploration of extraterrestrial environments reveals a wide array of unique conditions that challenge the very essence of life as we understand it. Adaptations to these conditions are not merely theoretical; they provide a framework for understanding how life could evolve in vastly different contexts across the universe. For instance, planets with extreme temperatures, high radiation levels, or unique atmospheric compositions necessitate innovative biological adaptations. Astrobiologists theorize that life forms on such planets may develop specialized proteins or cellular structures to withstand radiation or extreme heat, drawing parallels to extremophiles found on Earth, which thrive in harsh conditions.

One compelling aspect of extraterrestrial adaptations is the potential for different biochemical pathways. While life on Earth relies primarily on carbon-based molecules, the possibility of silicon-based life forms has been proposed for environments with limited carbon availability. These hypothetical organisms could exhibit entirely different metabolic processes, leading to diverse evolutionary pathways that reflect their unique surroundings. The study of these alternative biochemistries not only expands our understanding of life's possibilities but also raises questions about what constitutes an ecosystem and the potential interactions between various life forms on other planets.

Additionally, the social structures and cultures of extraterrestrial civilizations may also be shaped by their environmental conditions. For example, a species that evolved on a planet with scarce resources may develop communal living practices to ensure survival, while those on resource-rich worlds could adopt more individualistic or competitive strategies. The cultural implications of these adaptations are profound, influencing everything from social hierarchies to technological advancements. Understanding these dynamics offers insights into how extraterrestrial societies might function and interact, potentially providing a mirror to our own societal structures.

The interplay between environmental adaptations and government disclosure regarding extraterrestrial life adds another layer of complexity to the discourse. As new discoveries are made, and as evidence of extraterrestrial technology or civilizations emerges, the implications for our understanding of life beyond Earth grow increasingly significant. Conspiracy theories often arise from a lack of transparency in government communication about such discoveries, fueling speculation about the nature of extraterrestrial intelligence and its potential impact on human society. This intersection of science, culture, and conspiracy underscores the importance of open discourse in addressing questions about life in the cosmos.

Ultimately, the exploration of adaptations to unique extraterrestrial conditions invites us to reconsider our definitions of life and culture. The study of alien civilizations, their potential biochemistries, and the sociocultural ramifications of their environments challenges us to broaden our horizons and embrace the complexity of life beyond our planet. As we seek to understand the possibilities of extraterrestrial existence, the dialogue between science and speculation remains vital, guiding our quest to uncover the mysteries of the universe and our place within it.

5

Government Disclosure and Conspiracy Theories

Historical Context of UFO Disclosure

The historical context of UFO disclosure is a complex tapestry woven from threads of governmental secrecy, public fascination, and the evolving landscape of scientific inquiry. The modern era of UFO sightings began in the mid-20th century, coinciding with the onset of the Cold War, a time characterized by heightened security concerns and the proliferation of advanced technology. This period saw a significant uptick in reports of unidentified flying objects, many of which were documented by military personnel and civilians alike. The Roswell incident in 1947 became a focal point for conspiracy theories, igniting public interest and skepticism about what governments knew regarding extraterrestrial encounters.

Throughout the latter half of the 20th century, the United States government adopted a stance of denial and secrecy concerning UFOs. Programs such as Project Blue Book were established to investigate sightings, but their findings were often criticized for being inconclusive or misleading. This lack of transparency fostered a culture of distrust among UFO enthusiasts and conspiracy theorists, who believed that the government was concealing evidence of extraterrestrial life. As sightings continued and became more frequent, the narrative shifted from mere curiosity to a belief that these phenomena could represent a significant threat or offer profound insights into otherworldly civilizations.

The turn of the 21st century marked a notable shift in the discourse surrounding UFOs and government disclosure. High-profile military encounters, such as the 2004 Tic Tac incident involving U.S. Navy pilots, reignited public and scholarly interest in the phenomena. These incidents were captured on advanced surveillance systems and later confirmed by both military and intelligence officials, prompting calls for greater transparency. The establishment of the Unidentified Aerial Phenomena (UAP) Task Force in 2020 by the Pentagon represented a pivotal moment, signaling an official acknowledgment that warranted further investigation into these mysterious occurrences.

In parallel, the rise of social media and digital communication has transformed how information about UFOs is disseminated and discussed. Online platforms have become fertile ground for conspiratorial thinking, allowing individuals to share experiences, theories, and evidence in real time. This democratization of information has led to a more fragmented but passionate community of believers and skeptics alike, each armed with varying degrees of credibility. The intersection of digital culture and UFO phenomena has created a unique environment where historical events are continually reinterpreted, contributing to ongoing debates about the nature of extraterrestrial life and the implications of potential governmental cover-ups.

As we consider the historical context of UFO disclosure, it becomes clear that the journey toward transparency has been anything but linear. The interplay between government secrecy, public curiosity, and scientific inquiry has shaped the narrative surrounding unidentified aerial phenomena. With advancements in technology and a growing body of evidence, the conversation around UFOs is evolving. Whether one views these occurrences through the lens of skepticism or belief, they remain a compelling aspect of human culture, reflecting our enduring quest to understand our place in the cosmos and the possibility of life beyond Earth.

Major Government Reports and Their Implications

The exploration of extraterrestrial life and civilizations has long been a subject of intrigue, prompting numerous government reports that shape public perception and understanding. Key documents, such as the U.S. Department of Defense's Unidentified Aerial Phenomena (UAP) report, offer insights into the government's encounters with unidentified objects and the potential implications for national security. This report not only acknowledges the existence of phenomena that defy conventional explanations but also emphasizes the need for systematic investigation into these occurrences. The acknowledgment of UAPs by government entities has galvanized UFO believers and conspiracy theorists alike, fostering a dialogue surrounding the possibilities of extraterrestrial technologies and their implications for humanity.

Government reports often serve as a double-edged sword in the context of extraterrestrial research. While they can validate the experiences of individuals who have reported UFO sightings, they can also perpetuate skepticism among those who demand concrete evidence. The implications of these reports extend beyond mere acknowledgment; they challenge the scientific community to reassess the parameters of what constitutes valid research into extraterrestrial life. Astrobiologists, in particular, are prompted to explore unconventional avenues of inquiry that may lead to a deeper understanding of alien ecosystems and the potential for life beyond Earth. The existence of government-sanctioned investigations suggests a growing openness to the idea that humanity is not alone in the universe.

Furthermore, the revelations contained within government reports can have profound sociocultural implications. The public's reaction to these disclosures often reflects a broader existential curiosity about humanity's place in the cosmos. Reports that suggest the possibility of advanced extraterrestrial civilizations can inspire a range of responses, from fear and paranoia to excitement and hope. This cultural shift may lead to a reevaluation of philosophical and ethical frameworks as societies grapple with the implications of potential contact with intelligent life forms. Such perspectives are crucial for UFO believers and conspiracy theorists who often feel marginalized in their pursuit of understanding the universe.

In addition to shaping public perception, government reports influence policy and funding related to extraterrestrial research. The recognition of UAPs as significant phenomena has prompted calls for increased transparency and accountability from government agencies. This shift in policy can lead to enhanced collaboration between scientists and governmental bodies, paving the way for more rigorous studies into extraterrestrial life and civilizations. As funding for astrobiological research grows, it opens doors for innovative projects that may seek to uncover the mysteries of alien ecosystems and the cultural dynamics of potential extraterrestrial societies.

Ultimately, the implications of major government reports extend far beyond their immediate content. They serve as catalysts for debate and inquiry, encouraging a diverse array of stakeholders, from UFO enthusiasts to serious scientists, to engage with the complex questions surrounding extraterrestrial life. As these discussions evolve, they foster a deeper understanding of the potential interactions between human civilization and extraterrestrial cultures. The ongoing dialogue inspired by government disclosures will likely shape the trajectory of astrobiological research and the public's perception of our place in the universe for years to come.

Conspiracy Theories Surrounding Extraterrestrial Life

The fascination with extraterrestrial life has given rise to a plethora of conspiracy theories that often intertwine with claims of government cover-ups and secretive organizations. Many UFO believers assert that governments around the world, particularly the United States, possess classified information about alien encounters and technologies, deliberately withholding this knowledge from the public. This belief is fueled by historical incidents such as the Roswell crash of 1947, which ignited widespread speculation about alien spacecraft and government involvement in the retrieval of extraterrestrial beings. The narrative of a clandestine partnership between governmental bodies and alien civilizations continues to thrive, shaping the discourse around the existence of life beyond Earth.

Conspiracy theories surrounding extraterrestrial life often hinge on the concept of advanced civilizations engaging with humanity. Proponents suggest that these civilizations possess technologies far superior to our own, enabling them to traverse vast interstellar distances. The idea that extraterrestrial species are monitoring Earth or even influencing human development adds another layer to these theories. Some claim that ancient astronaut theories, which posit that aliens visited Earth in the distant past and shaped human cultures, reflect a deep-seated belief that humanity is not alone in the universe and has been in contact with these advanced beings for millennia. Such narratives challenge the traditional understanding of human history and suggest a more complex relationship with the cosmos.

Astrobiologists, while grounded in scientific inquiry, occasionally find themselves at the intersection of these conspiracy theories. The search for extraterrestrial life relies on the understanding of ecosystems and the potential for life to thrive in conditions vastly different from those on Earth. This scientific pursuit can inadvertently lend credibility to conspiratorial claims, as the possibility of life existing elsewhere fuels speculation about what those life forms might be like and how they could interact with humanity. Moreover, the discovery of extremophiles—organisms that thrive in extreme conditions—has expanded the scope of what constitutes a habitable environment, leading to further questions about the diversity of extraterrestrial ecosystems.

Government disclosure plays a pivotal role in the conspiracy narrative surrounding extraterrestrial life. The recent declassification of military reports on unidentified aerial phenomena has reignited public interest and speculation. Many conspiracy theorists argue that these disclosures are mere distractions or carefully crafted narratives designed to control the narrative surrounding alien encounters. The belief that more significant truths are hidden beneath the surface persists, prompting calls for transparency and accountability from governmental institutions. This ongoing dialogue emphasizes the tension between scientific exploration, public curiosity, and the often opaque nature of governmental operations.

In conclusion, the theories surrounding extraterrestrial life encompass a mix of genuine curiosity, speculative inquiry, and skepticism toward authority. As the search for extraterrestrial civilizations continues, the interplay between science and conspiracy will likely persist, influencing how individuals and societies understand their place in the universe. For UFO believers, conspiracy theorists, and astrobiologists alike, the quest for knowledge about alien species and their potential cultures remains an exhilarating endeavor, rife with possibilities and unanswered questions that challenge the boundaries of human understanding.

The Role of Whistleblowers in Alien Disclosure

Whistleblowers play a crucial role in the discourse surrounding alien disclosure, often serving as the linchpin between governmental secrecy and public awareness. These individuals, motivated by a sense of moral obligation, risk their careers and personal safety to unveil information that could reshape our understanding of extraterrestrial life and its implications for humanity. By coming forward with allegations of concealed evidence, secret research programs, or cover-ups regarding alien encounters, whistleblowers catalyze public interest and scholarly inquiry into the existence of extraterrestrial species and civilizations. Their testimonies can lend credibility to otherwise speculative narratives, creating a bridge between anecdotal experiences and the pursuit of scientific validation.

The impact of whistleblowers extends beyond mere revelations; they challenge prevailing paradigms and encourage a reevaluation of established beliefs. In the context of UFO sightings and encounters, whistleblowers have provided firsthand accounts that contradict official government narratives. By exposing discrepancies between what is publicly known and what is hidden, these individuals compel both the scientific community and the general populace to reconsider their stance on the possibility of extraterrestrial life. Such challenges to the status quo are essential in fostering an environment where open dialogue and rigorous investigation into alien phenomena can thrive.

Moreover, the motivations behind whistle-blowing in the realm of alien disclosure often stem from an ethical imperative to inform the public about potentially life-altering truths. Many whistleblowers express a deep concern for the implications that hidden knowledge could have on society at large. This sense of respon-sibility is particularly significant in discussions about advanced extraterrestrial technologies and their potential applications or dangers. By ex-posing classified information, whistleblowers not only advocate for transparency but also aim to ignite a collective conversation about the moral and ethical ramifications of interacting with alien civilizations.

The narratives put forth by whistleblowers of-ten intersect with conspiracy theories, adding layers of complexity to the discourse on alien life and governmental accountability. While some may dismiss these accounts as mere fabrications, others view them as vital pieces of a larger puzzle that reveals the extent of governmental secrecy and the potential existence of extraterrestrial ecosystems. This intersection of whistleblowing and conspiracy theory can mobilize communities of UFO believers and conspiracy theorists, fostering a collaborative effort to seek out the truth. Such alliances can lead to increased pressure on governments to disclose information, thereby contributing to a cultural shift toward greater acceptance of the possibility of extraterrestrial life.

In conclusion, the role of whistleblowers in alien disclosure is multifaceted, influencing public perception, scientific inquiry, and ethical considerations surrounding extraterrestrial encounters. Their courage to speak out against the backdrop of governmental secrecy not only enriches the narrative of alien existence but also underscores the importance of transparency in a democratic society. As the conversation around extraterrestrial civilizations and their cultures continues to evolve, the contributions of whistleblowers will remain pivotal in shaping our understanding of the cosmos and our place within it.

6

The Future of Galactic Societies

Implications of First Contact

The implications of first contact with extraterrestrial civilizations are profound and multifaceted, touching upon various aspects of human society, culture, and our understanding of the universe. The moment humanity establishes communication with an alien species will not only reshape our scientific paradigms but also fundamentally alter philosophical and religious beliefs that have persisted for centuries. As we grapple with the reality of intelligent life beyond Earth, questions regarding the nature of existence, our place in the cosmos, and the ethical responsibilities towards other sentient beings will emerge, demanding a reevaluation of long-held assumptions.

The sociopolitical landscape will inevitably be transformed by first contact. Governments around the world will face immense pressure to respond transparently to the revelations about extraterrestrial life. The dynamics of power may shift dramatically, as nations vie for leadership in interstellar diplomacy and cooperation. Conspiracy theories that have long circulated regarding government cover-ups will gain new momentum, with skeptics and believers alike questioning the validity of official narratives. This climate of suspicion could lead to increased activism for transparency and accountability, challenging the status quo in how information about extraterrestrial interactions is disseminated.

The cultural ramifications of first contact are equally significant. Alien civilizations may possess knowledge, art forms, and philosophies that challenge our understanding of creativity and existence. The exchange of ideas between species could lead to an unprecedented renaissance in human thought, fostering a spirit of collaboration that transcends national and cultural boundaries. However, there is also the risk of cultural imperialism, where dominant human cultures attempt to impose their values on alien societies. Striking a balance between mutual respect and the desire to learn from one another will be crucial in preserving the integrity of both human and extraterrestrial cultures.

In terms of ecological considerations, first contact could spark a renewed interest in the stewardship of our own planet. Understanding the ecosystems of alien worlds and their approaches to sustainability may provide valuable insights into our environmental crises. Moreover, the potential for interspecies cooperation in addressing galactic ecological concerns could emerge as a key aspect of our relationship with extraterrestrial civilizations. This collaboration may necessitate new ethical frameworks that prioritize the preservation of diverse ecosystems, both on Earth and beyond, expanding our understanding of ecological interconnectedness.

Finally, the prospect of first contact raises significant ethical questions that will challenge humanity to reconsider its moral obligation towards not just extraterrestrial beings, but also its own species. The ethical treatment of intelligent life forms, whether they are biologically similar or not, will demand a reevaluation of existing moral frameworks. The development of guidelines and principles for interstellar interactions will be essential to ensure that humanity acts as a responsible galactic citizen. As we stand on the brink of potential contact, the choices we make today will set the precedent for our future, influencing not only our relationship with extraterrestrial civilizations but also our legacy in the cosmos.

The Role of Humanity in a Galactic Community

The concept of humanity's role within a galactic community is a multifaceted inquiry that intersects with various disciplines, including astrobiology, sociology, and political science. As humanity ventures into the cosmos, the implications of our interactions with extraterrestrial civilizations become increasingly significant. The potential for collaboration, cultural exchange, and mutual understanding stands as a testament to the adaptability of human beings, underscoring our intrinsic desire to connect and communicate across vast distances. However, the complexities of these interactions are not merely theoretical; they require an examination of our ethical frameworks, technological capabilities, and the sociopolitical structures that govern our approach to extraterrestrial engagement.

In the context of extraterrestrial species and civilizations, humanity must grapple with the inherent diversity that characterizes the cosmos. Each civilization, influenced by its unique evolutionary path, environmental conditions, and cultural heritage, presents both challenges and opportunities for interaction. Human beings must cultivate an understanding of these differences, fostering an attitude of respect and curiosity rather than fear or domination. This perspective is crucial, as the potential for misunderstanding or conflict could arise from a lack of awareness and appreciation for the complexities of alien cultures. Engaging with these civilizations requires a willingness to learn and adapt, traits that have historically driven human progress.

Moreover, the role of humanity in a galactic community is not merely one of passive observation but active participation. As we develop the technological means to explore and possibly colonize other worlds, ethical considerations become paramount. Questions surrounding the preservation of extraterrestrial ecosystems and the rights of alien species must guide our endeavors. The responsibility to act as stewards of the cosmos, rather than conquerors, is essential for ensuring sustainable and equitable interactions. This stewardship encompasses a commitment to scientific inquiry, humanitarian principles, and the recognition of our shared existence within a broader galactic tapestry.

Government disclosure and conspiracy theories often shape public perception of humanity's role in the cosmos. The interplay between official narratives and alternative theories can complicate our understanding of extraterrestrial life and our position within the galactic community. Transparency regarding encounters with extraterrestrial civilizations is vital, as it influences not only scientific discourse but also the public's trust in institutions. By fostering an open dialogue about the existence of extraterrestrial species and the implications of such interactions, humanity can better prepare for the realities of engaging with the unknown.

Ultimately, the role of humanity in a galactic community hinges on our capacity for empathy, innovation, and ethical foresight. As we stand on the precipice of interstellar exploration, we must embrace a vision of coexistence that honors the rich diversity of life that may exist beyond our planet. The narrative of humanity must evolve from one of isolation to one of connection, recognizing that our actions in the cosmos will reverberate across the universe. Embracing this role requires a collective effort to understand, respect, and engage with the myriad forms of life that may share our galactic home, paving the way for a future defined by collaboration and mutual growth.

Ethical Considerations in Interstellar Relations

Ethical considerations in interstellar relations are paramount as humanity stands on the brink of potentially engaging with extraterrestrial civilizations. The prospect of encountering intelligent life beyond Earth raises complex moral questions that demand careful deliberation. The fundamental principle of respect for sentient beings, regardless of their origin, must guide our interactions. This respect extends to understanding their cultures, social structures, and technological capabilities, thereby ensuring that any engagement is based on mutual benefit and acknowledgment of their autonomy.

First and foremost, the ethical implications of communication with extraterrestrial species must be considered. Establishing contact carries the risk of cultural imperialism, where human values and norms could be imposed upon an alien society. This scenario is reminiscent of historical colonization on Earth, which often resulted in the exploitation and erasure of indigenous cultures. Thus, it is essential for humanity to approach interstellar communication with humility and a willingness to learn, recognizing that extraterrestrial civilizations may possess knowledge and philosophies that could significantly enrich our understanding of the universe.

Furthermore, the protection of extraterrestrial ecosystems should be a priority in interstellar relations. The introduction of human presence, whether through exploration or colonization, could disrupt delicate alien environments, leading to unforeseen consequences. Ethical stewardship requires a commitment to preserving these ecosystems, advocating for a non-intrusive approach to exploration that prioritizes the well-being of alien flora and fauna. This conservation ethic mirrors contemporary environmentalism on Earth, emphasizing the interconnectedness of all life forms and the responsibility we bear as stewards of our own planet.

Moreover, considerations surrounding technology transfer and its implications must be addressed. The sharing of advanced technologies with extraterrestrial civilizations presents both opportunities and risks. While such exchanges could foster collaboration and advancement, they could also lead to dependency or exacerbate inequalities between civilizations. It is crucial to navigate these technological exchanges with a focus on equitable partnerships, ensuring that both parties retain agency and benefit from the collaboration without compromising their core values or societal integrity.

Lastly, the issue of government disclosure and transparency becomes increasingly relevant in the context of interstellar relations. The public's right to know about potential extraterrestrial encounters is a fundamental ethical concern. Secrecy can fuel conspiracy theories and erode trust between governments and citizens. Ethical governance in the realm of interstellar affairs necessitates open communication, where information is shared responsibly, allowing for public discourse and engagement. The establishment of frameworks that prioritize transparency will not only enhance public understanding but also foster a sense of shared responsibility in navigating the complexities of interstellar interactions.

7

Understanding Our Place in the Universe

The Significance of Extraterrestrial Cultures

The exploration of extraterrestrial cultures is not merely an academic pursuit; it holds profound implications for our understanding of life beyond Earth. As we delve into the significance of these cultures, we recognize that they offer insights into the vast diversity of life forms that may exist in the universe. The study of extraterrestrial civilizations allows us to expand our definitions of culture, communication, and intelligence, challenging preconceived notions grounded in human experiences. These cultures could embody unique social structures, technological advancements, and philosophical outlooks that enrich our understanding of existence itself.

Examining extraterrestrial cultures also invites us to consider the potential for intergalactic communication and cooperation. If intelligent civilizations exist, the manner in which they might share knowledge and interact with one another could redefine our concepts of diplomacy and cultural exchange. The analysis of hypothetical or confirmed extraterrestrial societies opens avenues for understanding how advanced beings might approach issues such as conflict resolution, resource sharing, and environmental sustainability. Insights drawn from these cultures could inspire innovative solutions to terrestrial challenges, illustrating the interconnectedness of all life in the cosmos.

Moreover, the significance of extraterrestrial cultures extends to their ecosystems and the intricate relationships they maintain with their environments. Each civilization likely evolved within unique ecological contexts, leading to distinct cultural practices and belief systems shaped by their surroundings. Studying these ecosystems can provide valuable information about the adaptability and resilience of life, offering perspectives on how diverse life forms might respond to planetary changes. This knowledge could be crucial for astrobiologists seeking to identify potential habitable worlds and understand the parameters that support life.

The discourse surrounding extraterrestrial cultures also intersects with government disclosure and conspiracy theories, highlighting the societal impact of such revelations. The potential acknowledgment of intelligent extraterrestrial life by authorities could fundamentally alter humanity's understanding of its place in the universe. This shift would not only challenge established scientific paradigms but also provoke ethical considerations regarding our responsibilities to other sentient beings. The narratives surrounding these disclosures often reflect deep-seated fears and hopes, illustrating the complex relationship between knowledge, power, and belief.

In conclusion, the significance of extraterrestrial cultures transcends mere speculation, encapsulating vital themes about identity, coexistence, and the nature of life itself. As we continue to explore the possibilities of life beyond our planet, it becomes increasingly important to approach the subject with both scientific rigor and an open mind. Engaging with the potential realities of extraterrestrial civilizations not only enhances our understanding of the cosmos but also fosters a broader perspective on the diversity of life and the interconnectedness of all beings. This exploration ultimately challenges us to reflect on our own culture and civilization, encouraging a deeper appreciation for the myriad ways life can manifest across the universe.

Bridging the Gap Between Species

The concept of bridging the gap between species is a profound consideration in the context of interstellar relations and the potential for communication with extraterrestrial civilizations. The vastness of the universe and the diversity of life forms it may harbor necessitate a framework for understanding not only the biological differences among species but also the cultural and societal constructs that govern their interactions. For UFO believers and conspiracy theorists, the idea of contact with alien species raises questions about the nature of communication and the potential for cooperation or conflict. This subchapter explores the implications of such interactions and the methodologies that might facilitate understanding between diverse life forms.

To effectively bridge the gap between species, it is essential to recognize the limitations of human-centric perspectives. Human languages, customs, and social norms are deeply rooted in our evolutionary history and may not be applicable or relevant to extraterrestrial civilizations. Astrobiologists emphasize the need for a flexible approach to communication that transcends linguistic barriers. This might include the use of mathematics, a universal language of sorts, or the establishment of non-verbal communication methods such as visual symbols or electromagnetic signals. These methods would allow for a more inclusive dialogue with a variety of extraterrestrial species, each with its unique sensory perceptions and cognitive frameworks.

Moreover, understanding the ecosystems of extraterrestrial civilizations is crucial for fostering relationships. Each species evolves within its environment, developing specific adaptations and cultural practices that reflect their surroundings. Government disclosure regarding potential encounters with alien civilizations often focuses on the technological aspects of these encounters but overlooks the ecological and cultural dimensions. By studying the environmental contexts of various extraterrestrial species, we can gain insights into their values, social structures, and potential motivations, which are vital for establishing a foundation for mutual respect and collaboration.

The role of ethics in bridging the gap between species cannot be overstated. As we contemplate the prospect of engaging with extraterrestrial civilizations, ethical considerations must guide our interactions. This involves recognizing the sovereignty of other species and understanding the implications of our actions on their cultures and ecosystems. UFO believers and conspiracy theorists often highlight the fear and distrust surrounding government secrecy regarding extraterrestrial life. Establishing transparent and ethical frameworks for engagement would not only alleviate these concerns but also promote a sense of shared responsibility in the exploration of interstellar relations.

Ultimately, bridging the gap between species requires a commitment to understanding and respecting the diversity of life in the universe. It calls for collaboration across disciplines, including astrobiology, anthropology, and ethics, to develop comprehensive strategies for interaction. As we stand on the brink of potentially transformative encounters with extraterrestrial civilizations, it is imperative that we approach these possibilities with an open mind, a willingness to learn, and a dedication to fostering peaceful coexistence. In doing so, we may not only unravel the mysteries of the universe but also redefine our place within it.

The Ongoing Quest for Knowledge in the Cosmos

The quest for knowledge in the cosmos has become an increasingly significant pursuit, intertwining the disciplines of astrobiology, cultural anthropology, and the ongoing discourse surrounding government disclosures. As humanity ventures further into the universe, the search for extraterrestrial life and civilizations not only fuels scientific inquiry but also ignites the imaginations of UFO believers and conspiracy theorists. This multifaceted endeavor reflects a profound curiosity about the nature of existence, urging us to explore the possibility of intelligent life beyond our planet and the implications such discoveries would entail for our understanding of culture and society.

Astrobiologists have long posited that the conditions necessary for life might exist in the most unlikely of places. From the icy moons of Jupiter and Saturn to the exoplanets orbiting distant stars, the potential for diverse ecosystems challenges the traditional boundaries of biology and ecology. The ongoing search for microbial life in extreme environments on Earth provides a valuable framework for understanding how life might thrive in alien worlds. This research not only enhances our comprehension of life's resilience but also underscores the possibility of encountering extraterrestrial species with unique adaptations, complexities, and cultural frameworks.

The implications of discovering extraterrestrial civilizations extend beyond scientific inquiry, touching on philosophical, ethical, and sociopolitical considerations. The question of how these civilizations might communicate or interact with humanity looms large, raising issues about cultural exchange and the potential for conflict. Government disclosures regarding unidentified aerial phenomena (UAP) have fueled speculation about the existence of advanced extraterrestrial technologies and the possibility of covert interactions. Such narratives resonate deeply within conspiracy theories that challenge official accounts and suggest a deeper, hidden truth about our place in the cosmos.

Cultural exchanges between species could redefine our understanding of civilization itself. If we encounter extraterrestrial cultures, the potential for shared knowledge, art, and philosophy may pave the way for a new era of collaboration that transcends planetary boundaries. However, the risks associated with such encounters cannot be ignored. The history of human contact with indigenous cultures often serves as a cautionary tale, reminding us of the delicate balance between exploration and exploitation. It is essential to approach the prospect of interstellar dialogue with humility and respect, recognizing the rights and identities of other civilizations.

The ongoing quest for knowledge in the cosmos is an endeavor that unites diverse fields and perspectives. As we explore the mysteries of the universe, we must remain vigilant in our pursuit of truth, fostering an environment where scientific inquiry and open dialogue can flourish. The convergence of astrobiology, cultural studies, and the examination of government disclosures will undoubtedly shape our understanding of extraterrestrial life and its implications for humanity. In this grand quest, the most profound revelations may arise not only from what we discover but also from the questions we dare to ask as we navigate the uncharted territories of the cosmos.

www.ingramcontent.com/pod-product-compliance
Lightning Source LLC
Chambersburg PA
CBHW031210160726

47992CB00006B/2660